Just in Case

Saving Seeds in the Svalbard Global Seed Vault

MEGAN CLENDENAN *Illustrated by* **BRITTANY CICCHESE**

Charlesbridge

Only eight hundred miles from the North Pole,
away from wars and weapons,
safe from earthquakes, fire, and even an asteroid,
buried deep in the Earth on an island in Norway,
the Svalbard Global Seed Vault holds priceless treasure.

Steel doors open and beckon
into the belly of a mountain.

Inside the air is as cold as
the Arctic outside.
Walls of ice shine like stars.

The vault protects this treasure
in case of disaster . . .
in case of loss . . .
in case of war.

Just in case.

More than 580 million seeds, frozen.
Why and how did they get here?

For thousands of years,
our ancestors have planted seeds;
harvested, eaten, and sold what they grew;
and saved new seeds.
Generation by generation, kept for us today.

Seed Banks

There are about 1,700 seed banks around the world. Some, like the Millennium Seed Bank in England, are floodproof, bomb-proof, and temperature-controlled. Many others are small and simple community spaces with jars of seeds on shelves. All are vital to preserving future access to seeds and critical to crop biodiversity—a variety of plants in any given area. The Svalbard Global Seed Vault aims to store duplicates of every seed housed in all other seed banks.

Seeds are more than the foundation of our food.
They are history.
And they are the future.

Just like California condors, grizzly bears, and
monarch butterflies need care, so do seeds—
otherwise they could become extinct, gone forever.

The Hidden Secrets of Seeds

Just like there are many kinds of bears, birds, and butterflies, seeds have hundreds, if not thousands, of varieties. Each variety has a unique genetic code that determines the shape, size, and color of the plant that grows from the seed and also the type of environment where it flourishes. Some thrive where it's hot and dry. Others grow in chilly mountain areas. Some resist pests that others cannot.

We lose seeds when we stop using them.
Some farmers plant different varieties
and save seeds as they always have.

But many farmers now work with identical crops
that promise the biggest harvest.

Our planet's climate is changing.
Some types of seeds no longer grow where they used to.
We might need to plant different varieties in the future.

Facing Extinction

Even though there are more than 20,000 plants humans can eat, today we rely on only 12 major crops for the majority of our food. Of those, we use only a few varieties of seeds. For example, there are thousands of types of corn, but most corn in the United States is a variety called yellow dent. The unused varieties could eventually go extinct.

Seeds need a safe,
just in case.

On a faraway Arctic island of Norway,
architects, engineers, and scientists dreamed up a design.

No one had ever built a seed vault
in the middle of a frozen mountain.

POLAR BEARS SEEN NE

MOUNTAIN

UNDERGROUND

ENTRANCE

W
S → N
E

CONTROL ROOM

SEED
VAULTS

TUNNEL

SHELVES

TRØ450 to
LONGYEARBYEN = 1.5 HOURS

OSLO

Decisions, Decisions!

Should the vault be hidden or not? If the door is small, could it be obstructed by snow? If the vault is hidden with no road to get to it, how will seeds be transported? These questions and more were asked while planning the Svalbard vault. Final decisions included a location within binocular sight of the Longyearbyen airport tower.

TEMPERATURE:
-18°C

(BRR!)

To build the vault,
everything came from afar.

Machines—
excavators and diggers—
arrived by ship and plane,
enduring Arctic winds
and frozen seas.

Snow crunched
under work boots, and
breath fogged the air.

Diggers plowed through
white snow to reach
Earth's dark soil.

The mayor of
Longyearbyen, Svalbard,
sparked the first fuse,
a tiny flame that flashed and . . .

BLAST!

Time to Start Digging!

A ceremony was held in 2006, and construction got underway in April 2007. The first blast into the side of Plateau Mountain was on May 18 that year. The vault room is located more than 100 meters (328 feet) inside rock up to 60 meters (196 feet) thick. That's a lot of digging and carving!

Dust and smoke settled.
Solid rock was carved away
until a long tunnel hollowed the heart of the mountain.

Inside, orange-clad workers chiseled chamber walls
that rippled and rolled, a reminder that
this was a cave in a mountain.

Polar bears sniffed. They watched.

Puddles and Polar Bears

It was often cold and wet because the excavators melted the permafrost and created huge puddles. Polar bears roam Svalbard, and workers always had to keep watch for them. Arctic foxes are also common. One often appeared at lunchtime, and workers nicknamed it Salami.

Now metal that shines like silver coats the tunnel.

Copper pipes and black fans push cool air.

Sturdy concrete and heavy doors shield and shelter.

Rows of empty shelves wait.

A seed safe.

Just in case.

On Guard!

Svalbard Global Seed Vault keeps seeds safe
using different methods:
  motion, fire, and gas detectors,
  a series of five locked doors,
  temperature gauges,
  cameras,
  methane and radon detectors,
  and the local police force and coast guard.

Around the world, people prepare seeds
for the journey to the vault.

The packets that hold five hundred seeds each are tested.
Water,
heat,
air,
even stomping feet!
Nothing is a match for them.

Each Seed Has a Story

Countless people—farmers, gardeners, collectors all over the globe—worked hard to gather, dry, and send seeds to the vault. Each type of seed has a story that goes back generations. Someone saved them, named them, and grew food with them to feed their family. Each seed is important.

The first seeds land.
The Svalbard airport bustles with
planes and trucks and people.

Trolleys loaded with boxes
filled with seed packets
roll into the vault.

They pass through the tunnel
to a huge white room
where the cold lingers,
ice crystals shimmer, and
another set of locked doors opens.

Storing Seeds

The vault opened on February 26, 2008.
In the first five years, 80,000 unique crop
varieties arrived—but it wasn't an easy
journey! Seeds can't sit for long in warm
climates or on airplanes, so they must be
transported quickly. Once in the vault,
seeds are stored at -18°C (0°F). Because
of Svalbard's low temperatures and humidity,
some samples could potentially last hundreds
of years or more.

Inside the vault, a walk down the aisles becomes a walk through the world.

Seeds from
Pakistan and Peru,
Mongolia and Morocco,
North Korea and South Korea.
Seeds from almost
every country.

The First Withdrawal

In 2011 a war raged in Syria. A seed bank located in that country, full of lentils, chickpeas, and other important crops, was in danger. Before people fled, they sent 116,000 varieties of seeds to the Svalbard vault. Four years later, 38,703 varieties were retrieved—the first withdrawal. Those seeds were planted in Lebanon and Morocco, where they now thrive. New seeds were returned to the Svalbard vault, a full circle.

A treasure trove of
tiny kernels of life,
that if planted
will sprout,
soak up the sun,
drink the rain,
provide oxygen,
and nourish us.

More than 580 million seeds.
For you, for me,
for everybody.

Just in case.

KEEPING SEEDS IS DIFFICULT

There are about 1,700 seed banks around the world. Keeping seeds safe is not easy. During World War II, one of the oldest and largest seed collections in the world, the Vavilov Institute of Plant Industry in Saint Petersburg, Russia, faced invasion from Nazi Germany. People risked and lost their lives to save the seeds. Wars in Iraq, Afghanistan, and Burundi have also caused significant loss of seed collections. In 2022 in Ukraine, thousands of seeds from the national collection were destroyed by a Russian bomb. Many varieties were not duplicated anywhere else yet.

Even in peacetime, seeds are not always safe. In 2004 in Bari, Italy, the seed bank freezers failed during a heat wave, and repairs took months to complete, so seeds were lost. In 2006 in the Philippines, a typhoon destroyed a seed bank. In the 1990s an evaluation of the world's seed banks found that many locations weren't secure. Some didn't have enough money to keep equipment working or the power on. Some were too hot, others too moist. Something needed to be done before more seeds were lost forever.

After unseasonably warm Arctic temperatures resulted in rain instead of snow in early 2017, part of the entry of the Svalbard Vault flooded. No seeds were damaged, and the tunnel was renovated to ensure it is fully waterproof.

A SOLUTION FOR SEED SAFETY

The Svalbard Global Seed Vault was built as a backup for the other seed banks in the world. Duplicates of seeds in other vaults are also stored at Svalbard. Each year more seeds are deposited. They represent 12,000 years of history among people, plants, and the environment. Seeds are what have sustained generations of humans, allowing us to grow the food we need to live. The plan is that if any seed collection is destroyed by war, natural disaster, or poor management, copies will be at Svalbard. Just in case.

A group of conservationists considered the best location to build this secure seed vault. Svalbard, a Norwegian island, offers solid stone mountains protected from war, weather, rising seas, and earthquakes. It is remote and naturally cold, which helps protect seeds from equipment failure. Norway has a reputation for being a fair and safe country that other countries would trust to safeguard their seeds.

The group presented their idea to the Norwegian government and asked for support. They got it! Although the vault serves the whole world, Norway paid to construct it. It is owned by Norway and managed with support from the Nordic Genetic Resource Center (NordGen) and the Crop Trust, an international nonprofit organization dedicated to preserving crop diversity for food security. Norwegian artist Dyveke Sanne designed a sculpture for the entrance to the seed vault made of pieces of polished steel that reflect light and glow greenish-turquoise in the dark polar night.

Each packet sent to the vault holds 500 seeds, copies rather than the only ones of each variety. The people who send seeds still own them. The vault opens a few times each year for deposits, and like a regular bank, only the depositor can access their seeds. The hope is to save each seed's genetic code for the future.

Farmers today need to produce more food on less land because there are more people to feed. Climate change means our food crops must adapt to higher temperatures, longer periods of hot weather, extreme rains, and changing migrations of pests and pollinators.

By saving varieties of seeds, we can help adapt our food for the future. For example, we might need to plant a variety of wheat that can grow with less water. The best-growing wheat, rice, and corn today may not grow as well or at all in the future. But where will we find the new wheat, rice, and corn? They will be part of the diversity of seeds, either growing in a field or saved in a seed bank.

Some estimates say that about 75 percent of seed crops have been lost in the last 125 years. There's no more time to lose to save the foundation of our food.

WHAT'S IN THE VAULT?

As of 2023, the Svalbard Global Seed Vault protects samples of more than 1.2 million seed varieties. And it has the capacity to store 4.5 million varieties. With 500 seeds per packet, that's about 2.25 billion total seeds.

Surrounded by the frozen mountain fortress and cold dry air, some seeds can be stored for decades, others possibly hundreds of years. But long-term seed storage still isn't perfect. If the seeds aren't sometimes replanted and replaced, they may no longer be suited to the climate where they originally grew best because they won't evolve with environmental changes. Seed banks at community or country levels can test seeds to see which need to be rejuvenated, and then redeposit into Svalbard.

The total number of different types of seeds in the vault changes as more seeds are deposited. For example, in 2022 there were about 80,000 samples of barley in the vault. A year later in 2023, there were 110,000. The vault also includes ever-growing numbers of other major crops such as wheat, rice, sorghum, beans, maize, soybeans, and chickpeas.

Plant a Seed, Save a Seed

The best way to save seeds for the future is to grow them yourself! Though seed vaults act as an ultimate backup plan, living conservation happens when farmers and gardeners around the world plant, harvest, and replant seeds the following year. Seeds can be saved anywhere by anyone—even you! Enjoy what you grow. Save the new seeds for next time. Just in case.

AUTHOR'S NOTE

I first learned about the Svalbard Global Seed Bank from a podcast. I was astounded that we built a vault in a mountain to protect our seeds! Even though I have grown many vegetables since I was a child, I never gave much thought to the security of seeds or crop biodiversity.

I read books, academic articles, and websites to learn more. I watched documentaries and TED Talks. I interviewed Dr. Cary Fowler, one of the people who was deeply influential in creating the Svalbard Global Seed Vault.

The more I learned, the more in awe I became about how many people save seeds for our future: farmers, gardeners, scientists, and so many more. I hope you're inspired to learn more about seed saving, crop biodiversity, farming practices, and the future of our food.

Check at your local library—they might have a "seed library" where you can borrow and exchange local seeds.

RESOURCES FOR KIDS

Learn how to plant a garden from seeds.

Archer, Joe, and Caroline Craig. *Plant, Cook, Eat! A Children's Cookbook.* Watertown, MA: Charlesbridge, 2018.

Learn about different kinds of farmers and crops around the world.

Castaldo, Nancy, and Ginnie Hsu. *The World That Feeds Us.* Beverly, MA: Quarto Group USA, 2023.

Learn about how corn has changed over time and why it's important around the world.

Piñeyro-Nelson, Alma, Daniela Sosa-Peredo, Emmanuel González-Ortega, and Elena R. Álvarez-Buylla. "There Is More to Corn than Popcorn and Corn on the Cob!" Frontiers for Young Minds, December 13, 2017. https://kids.frontiersin.org/articles/10.3389/frym.2017.00064.

Take a virtual tour of the Svalbard Seed Vault.

https://www.croptrust.org/work/svalbard-global-seed-vault

Read about the vault and see charts of the countries that have seeds stored there.

https://seedvault.nordgen.org

SELECTED BIBLIOGRAPHY

Books

Fowler, Cary. *Seeds on Ice: Svalbard and the Global Seed Vault.* Westport, CT: Prospecta Press, 2016.

Nabhan, Gary Paul. *Where Our Food Comes From: Retracing Nikolay Vavilov's Quest to End Famine.* Washington, DC: Island Press, 2008.

Saladino, Dan. *Eating to Extinction: The World's Rarest Foods and Why We Need to Save Them.* New York City: Farrar, Straus and Giroux, 2022.

Websites

Svalbard Global Seed Vault, www.seedvault.no

Crop Trust, www.croptrust.org

Seed Savers, www.seedsavers.org

United Nations Convention on Biological Diversity, www.cbd.int/agro

Interview

Fowler, Cary. Personal interview with the author, February 25, 2022.

FOR DAD, WHO GAVE ME MY FIRST SEEDS TO PLANT
—M. C.

FOR THE SCIENTISTS AND FARMERS WORKING TOWARD A MORE SUSTAINABLE FUTURE
—B. C.

Text copyright © 2025 by Megan Clendenan
Illustrations copyright © 2025 by Brittany Cicchese
All rights reserved, including the right of reproduction in whole or in part in any form. Charlesbridge and colophon are registered trademarks of Charlesbridge Publishing, Inc.

At publication, all URLs in this book were accurate. Charlesbridge, the author, and the illustrator are not responsible for the content of any website.

Charlesbridge • 9 Galen Street, Watertown, MA 02472 • www.charlesbridge.com

Library of Congress Cataloging-in-Publication Data
Names: Clendenan, Megan, 1977– author. | Cicchese, Brittany, illustrator.
Title: Just in case: saving seeds in the Svalbard Global Seed Vault / Megan Clendenan, illustrated by Brittany Cicchese.
Other titles: Saving seeds in the Svalbard Global Seed Vault
Description: Watertown, MA: Charlesbridge, [2025] | Includes bibliographical references. | Audience: Ages 5–8 | Audience: Grades 2–3 | Summary: "Deep in the mountains of Norway, the Svalbard Seed Bank stores seeds of the world's food supply."—Provided by publisher.
Identifiers: LCCN 2024031590 (print) | LCCN 2024031591 (ebook) | ISBN 9781623544805 (hardcover) | ISBN 9781632894236 (ebook)
Subjects: LCSH: Svalbard Global Seed Vault. | Seeds—Preservation—Juvenile literature. | Seed supply—Juvenile literature. | Germplasm resources conservation—Juvenile literature.
Classification: LCC SB118.38 .C54 2025 (print) | LCC SB118.38 (ebook) | DDC 631.5/21—dc23/eng/20250107
LC record available at https://lccn.loc.gov/2024031590
LC ebook record available at https://lccn.loc.gov/2024031591

Printed in China • OPIC
(hc) 10 9 8 7 6 5 4 3 2 1

Illustrations created digitally
Text type set in Neutraface
Edited by Karen Boss
Designed by Kristen Nobles
Production supervised by Jennifer Most Delaney